A World of Birds

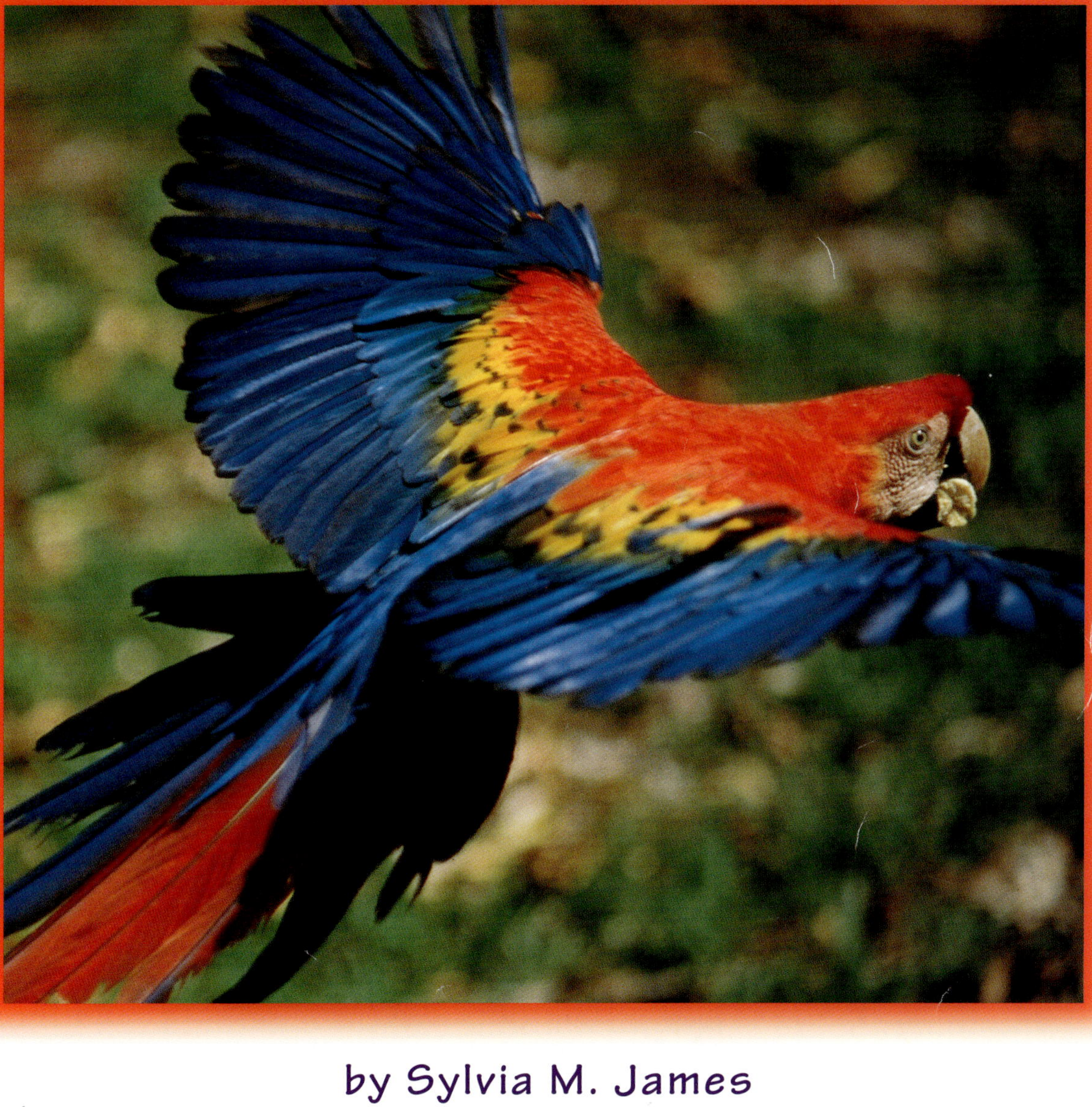

by Sylvia M. James

MONDO

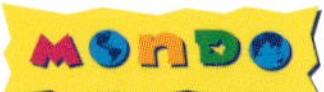

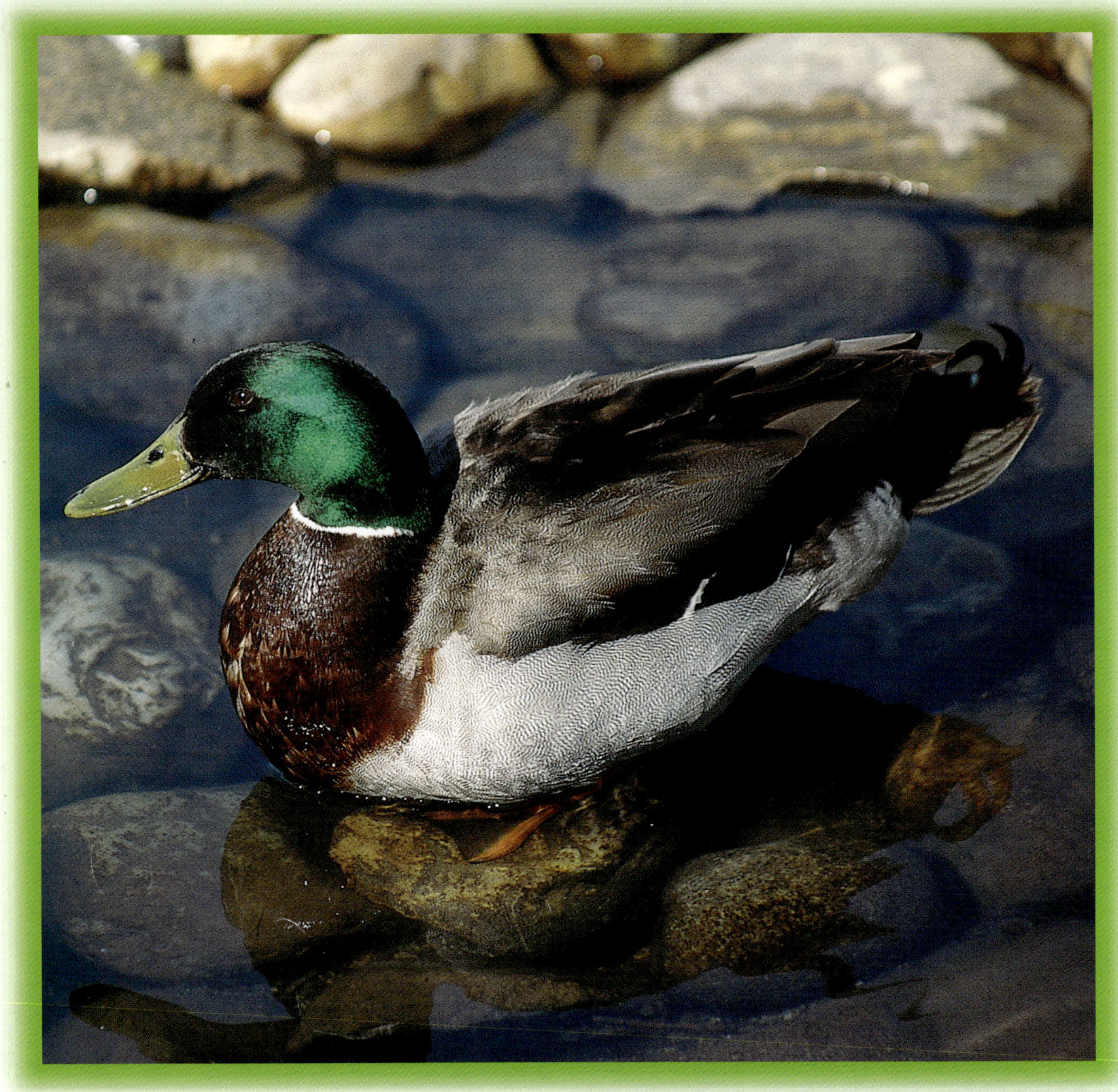

See this bird. It is a duck.

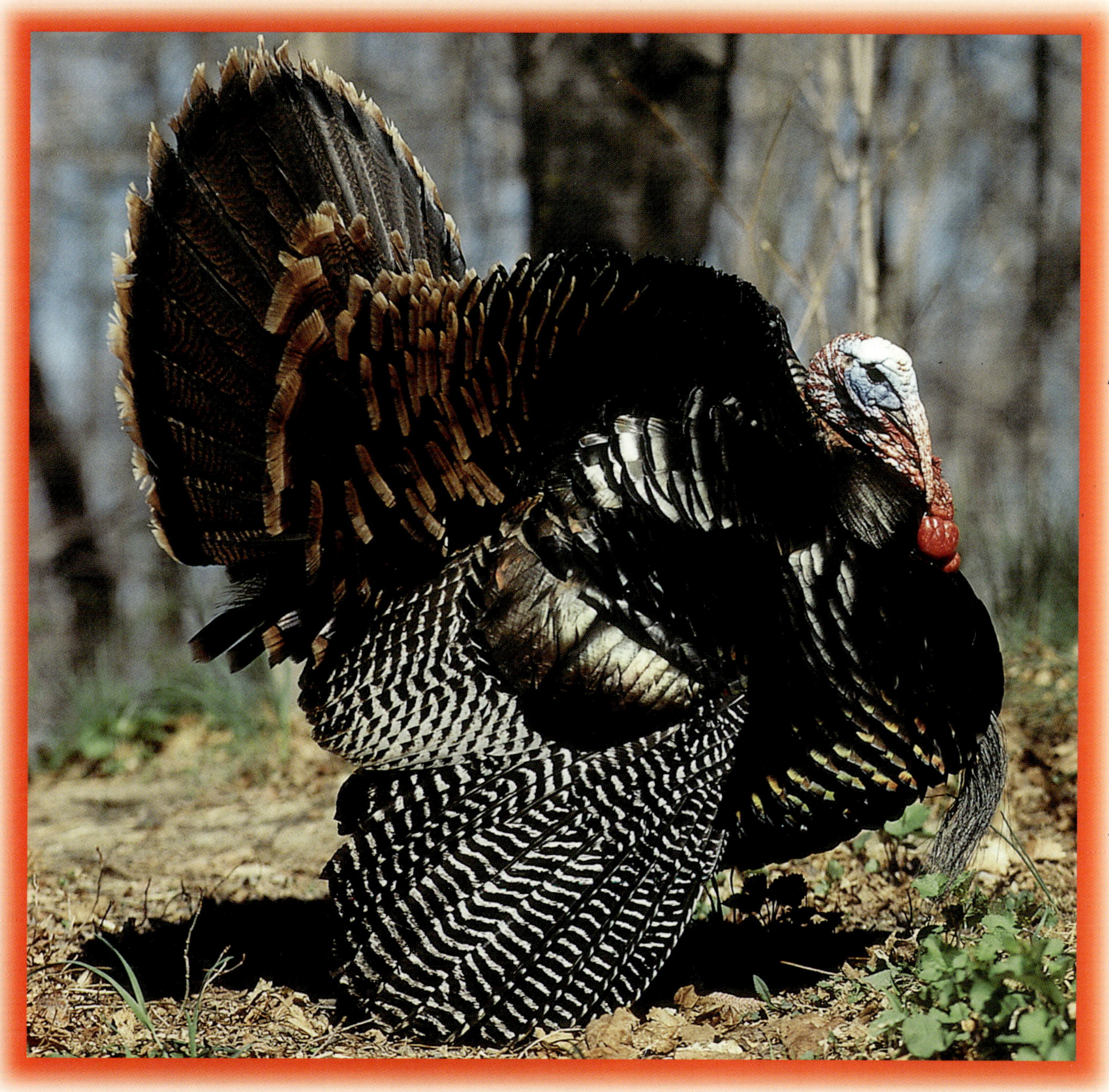

See this bird. It is a turkey.

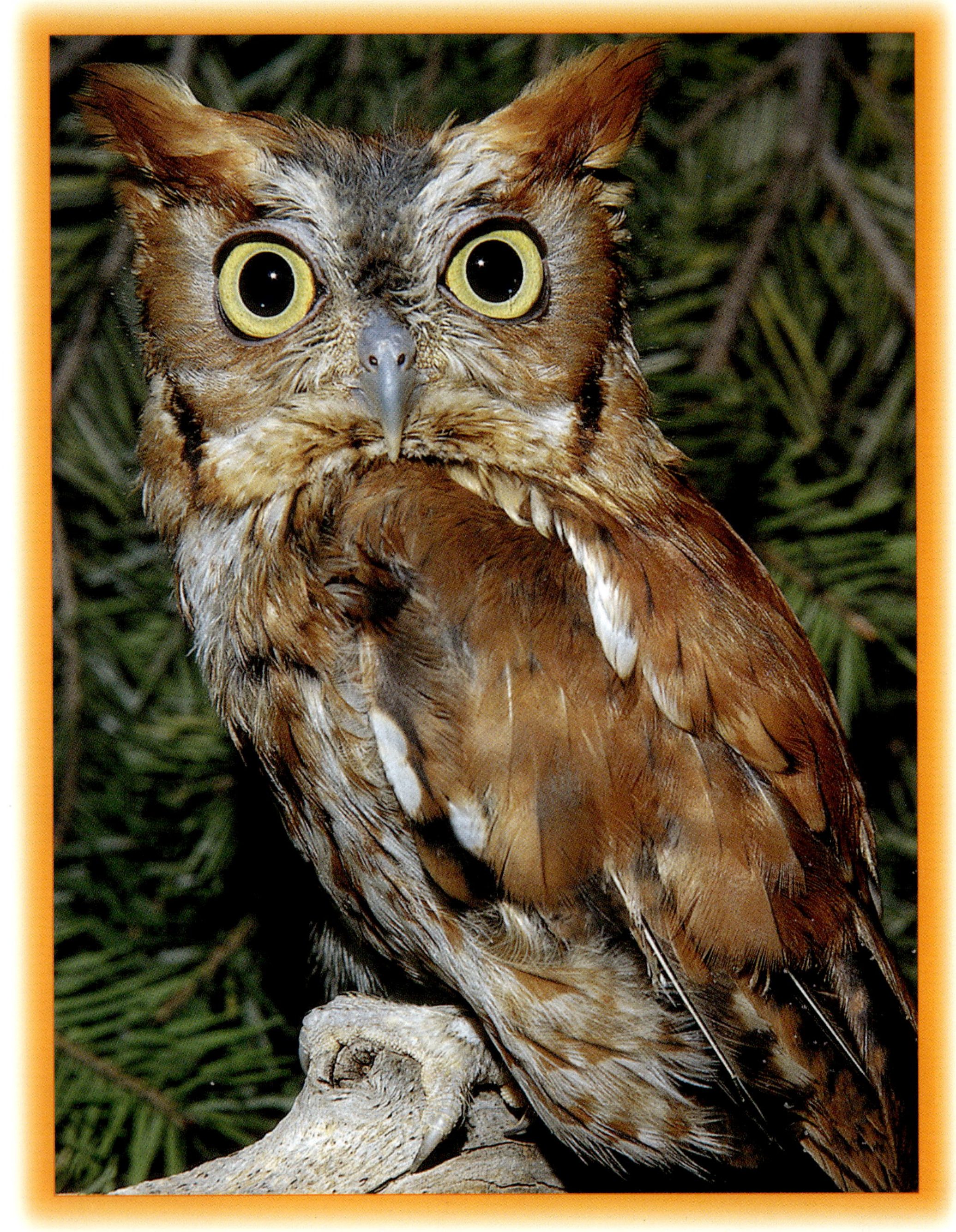

See this bird. It is an owl.

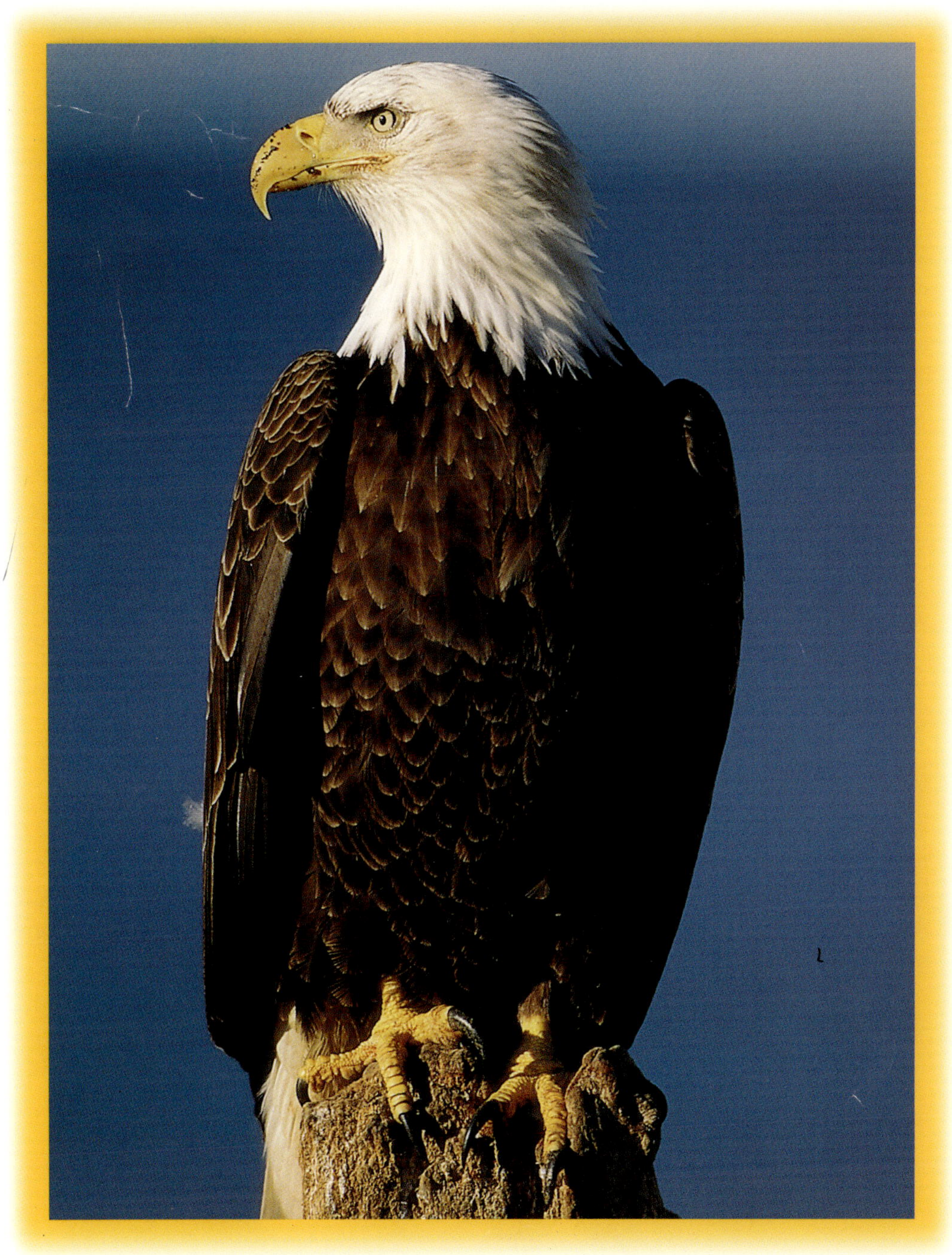

See this bird. It is an eagle.

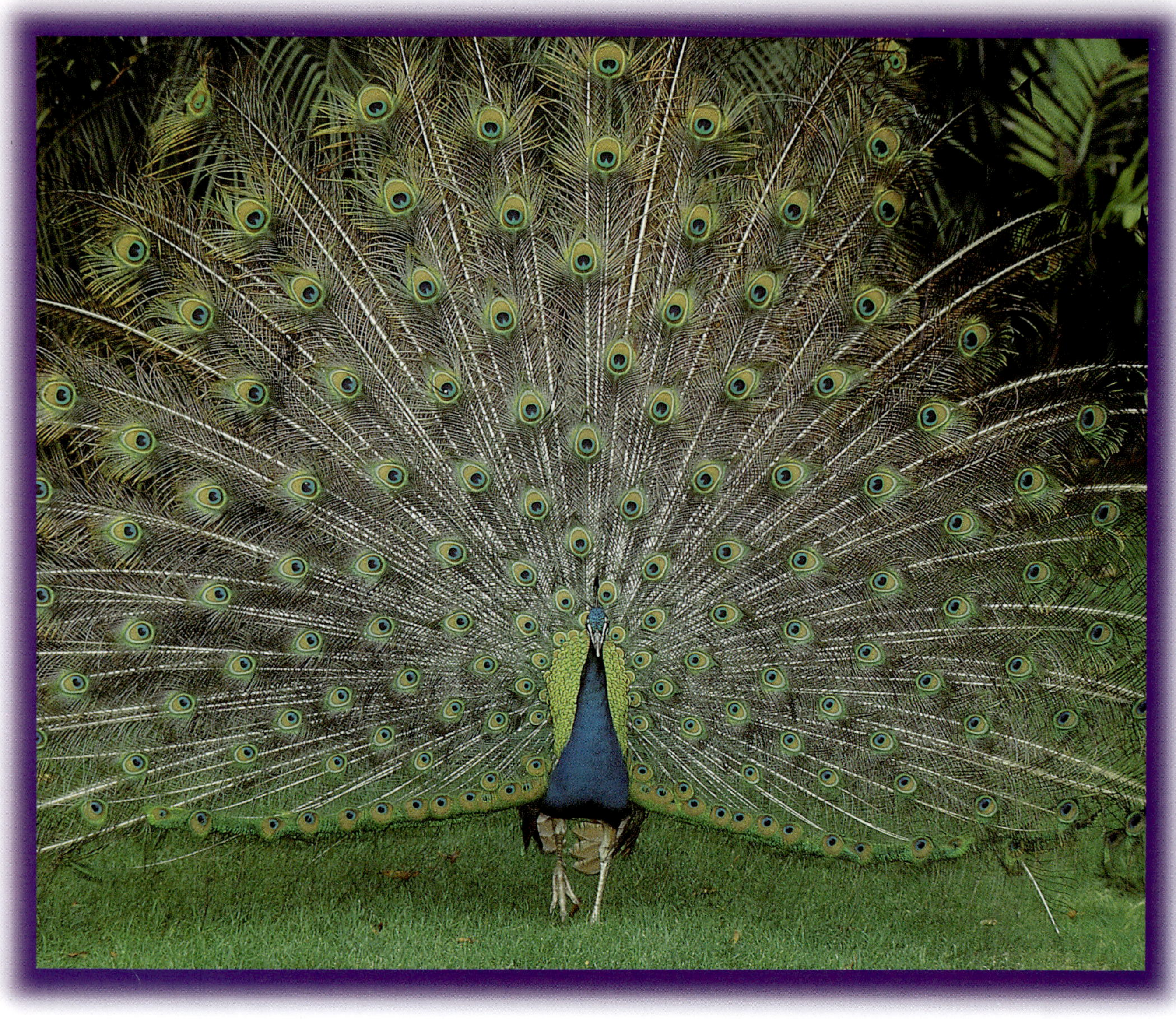

See this bird. It is a peacock.

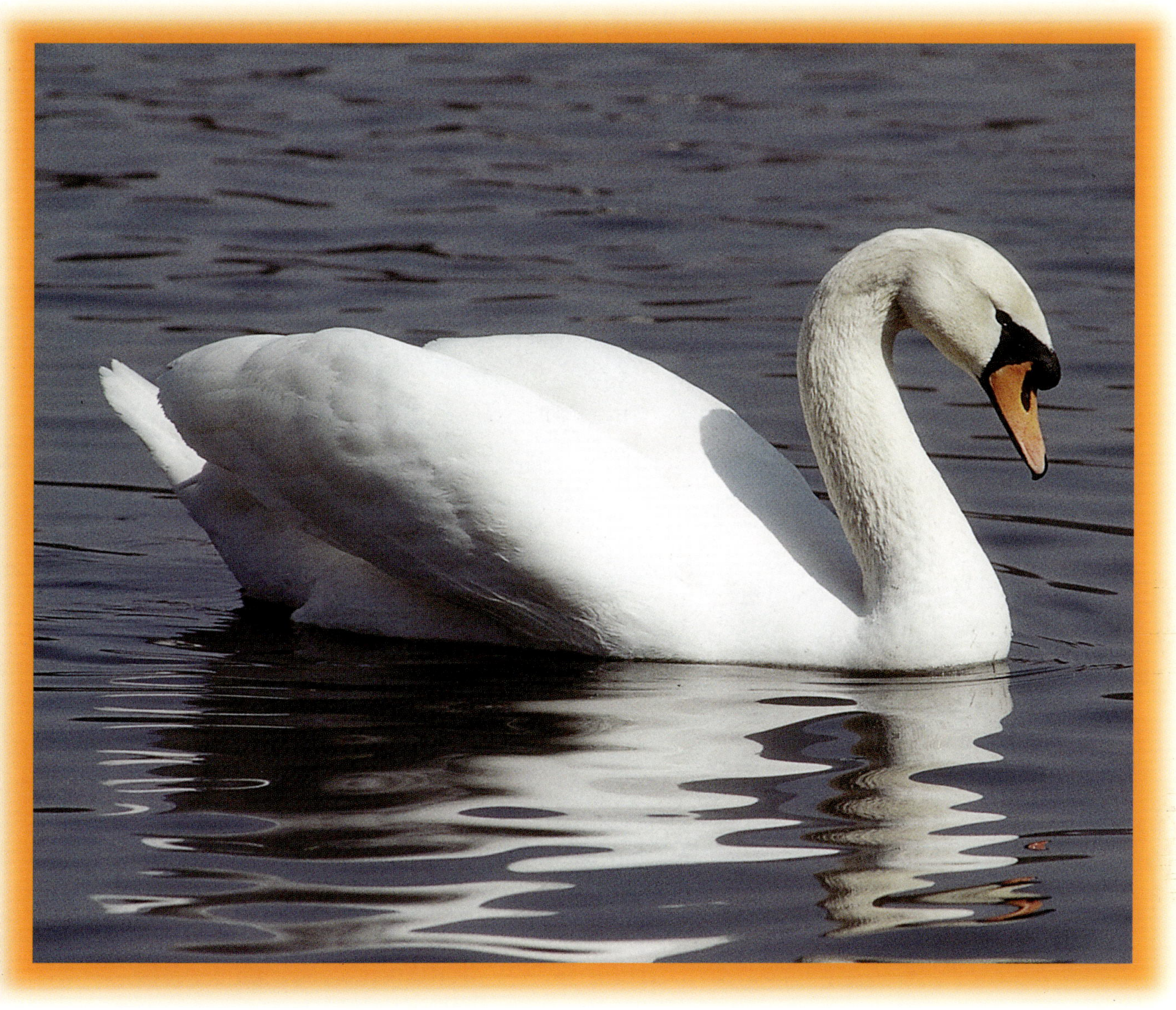

See this bird. It is a swan.

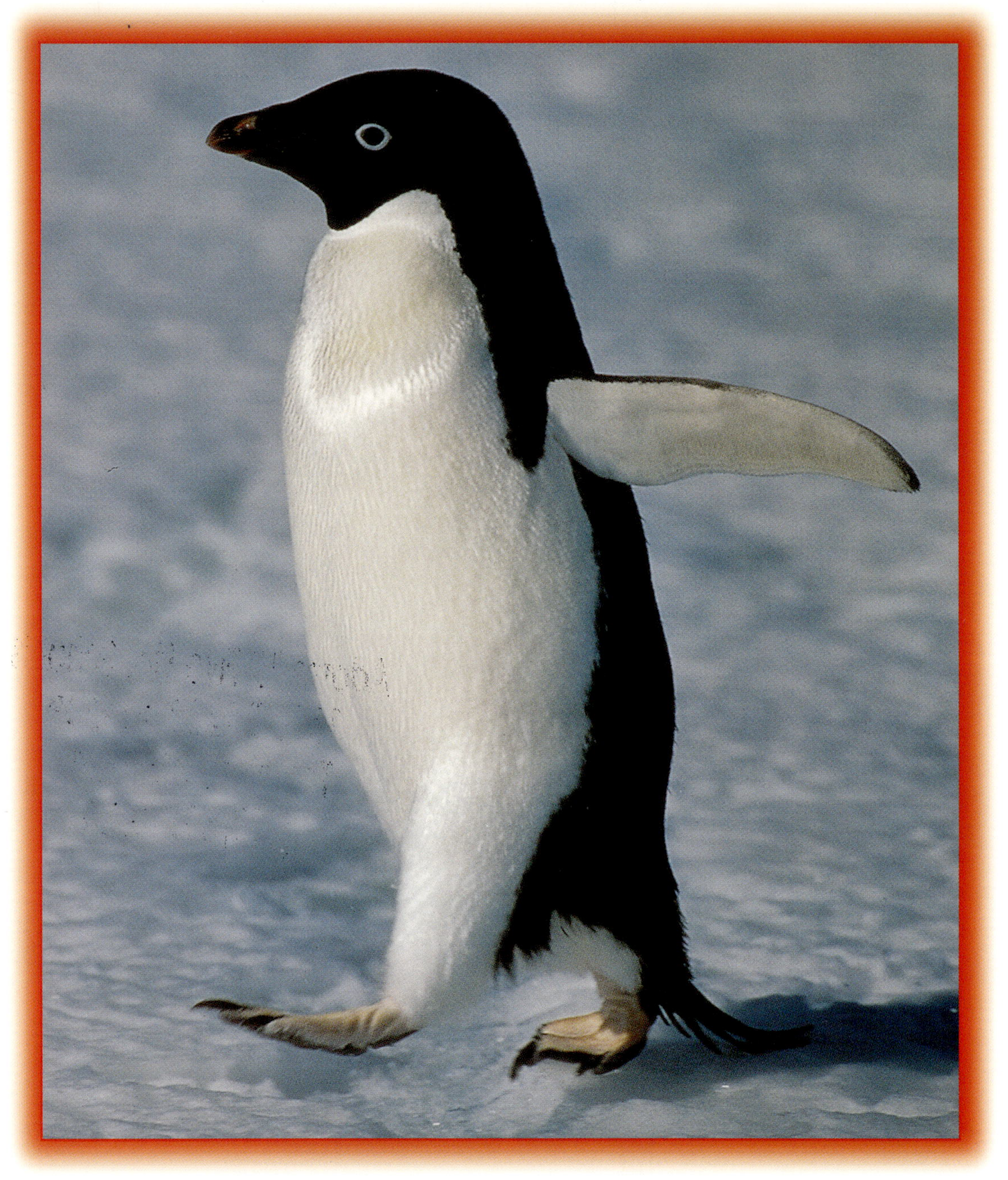

See this bird. It is a penguin.